## NOTICE

This report was prepared as an account of work sponsored by an agency of the United States government. Neither the United States government nor any agency thereof, nor any of their employees, makes any warranty, express or implied, or assumes any legal liability or responsibility for the accuracy, completeness, or usefulness of any information, apparatus, product, or process disclosed, or represents that its use would not infringe privately owned rights. Reference herein to any specific commercial product, process, or service by trade name, trademark, manufacturer, or otherwise does not necessarily constitute or imply its endorsement, recommendation, or favoring by the United States government or any agency thereof. The views and opinions of authors expressed herein do not necessarily state or reflect those of the United States government or any agency thereof.

Available electronically at http://www.osti.gov/bridge

Available for a processing fee to U.S. Department of Energy
and its contractors, in paper, from:
U.S. Department of Energy
Office of Scientific and Technical Information
P.O. Box 62
Oak Ridge, TN 37831-0062
phone: 865.576.8401
fax: 865.576.5728
email: mailto:reports@adonis.osti.gov

Available for sale to the public, in paper, from:
U.S. Department of Commerce
National Technical Information Service
5285 Port Royal Road
Springfield, VA 22161
phone: 800.553.6847
fax: 703.605.6900
email: orders@ntis.fedworld.gov
online ordering: http://www.ntis.gov/ordering.htm

**Assessing and Improving the Accuracy of Energy Analysis for Residential Buildings**

Prepared for:

Building America

Building Technologies Program

Office of Energy Efficiency and Renewable Energy

U.S. Department of Energy

Prepared by:

Residential Buildings Research Group

Electricity, Resources & Building Systems Integration Center

National Renewable Energy Laboratory

1617 Cole Boulevard

Golden, CO 80401

July 2011

# Acknowledgments

This work was funded by the U.S. Department of Energy (DOE) Building Technologies Program. The authors wish to thank David Lee (DOE Team Leader, Residential Buildings) for his continued support. We would also like to thank Joel Neymark of J. Neymark & Associates, as well as Ren Anderson, Marcus Bianchi, Jay Burch, Craig Christensen, and Ron Judkoff, all of the National Renewable Energy Laboratory, for their valuable guidance and feedback.

# Nomenclature

ASHRAE        American Society of Heating, Refrigerating and Air-Conditioning Engineers

BESTEST       Building Energy Simulation Test

$CFM_{50}$    cubic feet per minute at a pressure difference of 50 Pascals

DOE           U.S. Department of Energy

$e$           error

$\bar{e}$     mean error

HERS          Home Energy Rating System

HES           Home Energy Saver

$m$           measured value

NREL          National Renewable Energy Laboratory

NRMSE         normalized root mean squared error

ORNL          Oak Ridge National Laboratory

$p$           predicted value

RMSE          root mean squared error

# Contents

# List of Figures

# List of Tables

# 1 Introduction

Whole-building energy analysis is used in the residential sector for many purposes:

- Design energy-efficient homes.

- Produce labels, scores, and ratings.

- Predict energy and cost savings from energy efficiency upgrades (retrofit measures).

- Determine cost and performance criteria for new energy-efficiency technologies.

- Provide quantitative analysis and data to support programmatic and policy-related decisions.

The success of energy-efficient design, labeling, scoring, rating, and retrofit efforts depends largely on the accuracy of the analysis performed for each task; stakeholders must be confident that the analysis approach can accurately predict relevant metrics such as energy use and energy savings. This report describes the National Renewable Energy Laboratory's (NREL) methodology to assess and improve the accuracy of whole-building energy analysis for residential buildings.

## 1.1 Mechanisms To Improve Energy Analysis Methods

A variety of approaches, including annual building energy simulation, statistical analysis based on empirical data, and spreadsheet calculations, can be used to perform whole-building energy analysis. This paper focuses on improving whole-building energy analysis methods that employ annual building energy simulations. These methods involve inputting information about a building into a building energy simulation program and running the program to predict energy use. Key participants are energy assessors, building energy analysts, and software developers. NREL's improvements to the accuracy of energy analysis methods are translated to industry through three primary mechanisms:

- **Field data collection procedures.** Published recommendations for collecting data from a house through measurement and observation.

- **Simulation protocols.** Published defaults and assumptions for simulation inputs that are not directly measured or observed in the field.

- **Software test suites.** Published suites of tests through which commercial vendors can compare their software results to those of reference software,[1] analytical solutions, and empirical data. Tests can be at the whole-building or isolated-component level.

Figure 1 shows how each mechanism relates to energy analysis in the field. This report describes how NREL's overall approach for improving the accuracy of energy analysis methods for residential buildings leads to improvements in field data collection procedures, simulation protocols, and software test suites. The approach builds on previous work at NREL to develop a methodological foundation for evaluating the accuracy of whole-building energy simulation programs (ASHRAE 2007; Judkoff and Neymark 1995a, 1995b; Judkoff et al. 2010; Neymark

---

[1] Detailed, hourly whole-building energy simulation engines such as EnergyPlus, DOE-2, and SUNREL®.

and Judkoff 2004, 2008, 2009a), but also focuses on assessing and improving the accuracy of input data for residential energy analysis methods.

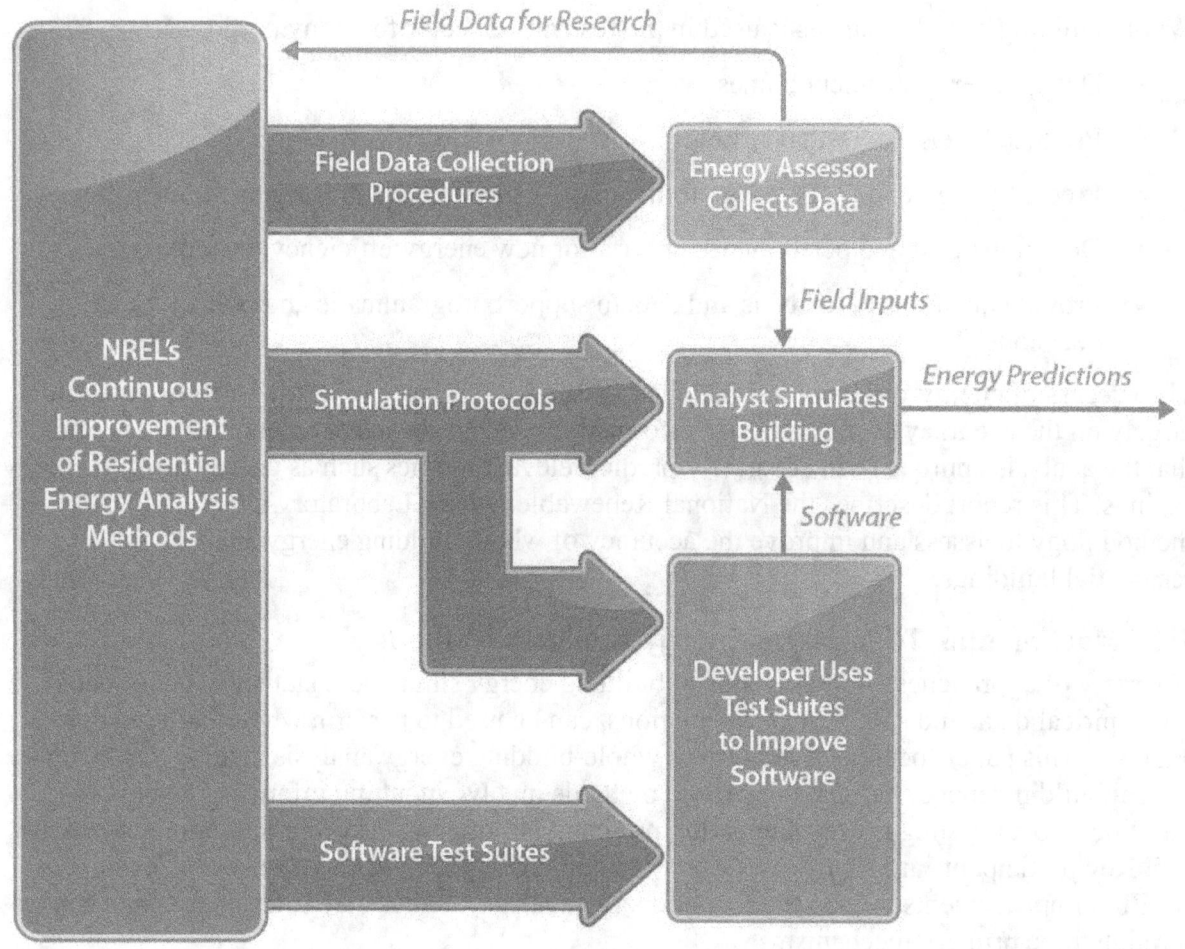

**Figure 1. Mechanisms for improving energy analysis methods in the field**

# 2 Motivation

Historically, DOE/NREL residential buildings research efforts have focused on high-efficiency designs for new homes. Software has been validated primarily in the context of such buildings. During the past few years, more emphasis has been placed on research related to improving the efficiency of existing homes. Anecdotal evidence and controlled studies have raised concerns about the accuracy of software-based energy analysis for existing homes, especially for energy-inefficient ones.

A general perception is that software-based energy analysis of inefficient existing homes tends to overpredict pre-retrofit energy use and retrofit energy savings. This perception hinders the efforts of companies, utilities, programs, and research institutions that depend on the accuracy of energy analysis predictions. One goal of our overall research is to use empirical data from the field to assess and track the accuracy of reference software. A literature review of previous accuracy studies is presented in Appendix A, from which we found that:

- Multiple studies confirm that analysis methods tend to overpredict energy use and savings in poorly insulated, leaky homes with older mechanical systems, that is, homes most needing energy retrofits.

- Determining the reason for discrepancies is difficult, because all error sources act simultaneously.

- Many metrics for accuracy are used in various studies and many possible causes of overprediction are described. A single reference containing definitions for error, potential sources of error, and accuracy, developed in the context of comparing simulated building energy use to measured energy use, would be beneficial.

- A method to improve the accuracy of energy use and savings predictions is needed.

Definitions for error, sources of error, and accuracy are presented in Appendix B. These can be used to improve the consistency of calculations for errors and accuracy, which provides for easier comparison across studies and reduces confusion that arises from having differing definitions in different sources. The example metrics presented in Table 1 comprise an initial list; more metrics may be added as efforts to improve the accuracy of analysis methods continue.

As described in Appendix B, sources of error can be grouped into the following categories:

- Related to the analysis methods
  - o Inputs
  - o Software.
- External to the analysis methods
  - o Processing utility data
  - o Quality and consistency of retrofit contractor work.

The method described in this report focuses on reducing errors related to the analysis methods (inputs and software)—methods for reducing error external to the analysis methods are not described. A method for improving the accuracy of energy use and savings predictions is described in Section 3.

# 3 Method to Improve the Accuracy of Residential Energy Analysis Methods

A continuous process for improving the accuracy of energy analysis for residential buildings is illustrated in Figure 2 and is described in Sections 3.1–3.7. The first four subprocesses (Sections 3.1–3.4) reduce software and input errors through improvements to field data collection procedures, simulation protocols, and software test suites. The last three (Sections 3.5–3.7) track the impact of improving inputs and reference software by comparing analysis predictions to metered energy use in real homes.

## 3.1 Identify Potential Issues

The method illustrated in Figure 2 involves incremental improvements to energy analysis methods by identifying, investigating, and correcting specific input and software issues. The first step is to identify potential issues based on:

- **Comparisons of predicted versus metered energy use and savings.** With large samples of predicted and metered energy usages, observation and statistical analysis of errors may identify certain building characteristics, occupant types, and building sites where energy analysis methods perform poorly.

- **External methods.** Hypotheses about specific issues may originate from consensus of modeling experts, field tests, comparative software tests, etc.

Issues with the greatest potential impact on energy use and savings predictions, as determined from the literature and preliminary sensitivity and uncertainty studies, are given priority. A selected issue is typically classified as either an input issue or a software issue. In some cases the classification may be difficult to determine. For example, if the potential issue is *heat transfer through uninsulated wall cavities,* one could hypothesize that (1) the default input for the air-gap thermal resistance is incorrect; or (2) the current wall heat transfer model in the reference software is not appropriate for uninsulated cavities. Because models define the necessary input values (in this example the 1-D conduction model in the software defines the need to input an air gap thermal resistance), the best approach is to first confirm the model is accurate and then verify that methods for defaulting, estimating, or measuring the relevant inputs are sufficient.

## 3.2 Improve Inputs

Potential input issues (see Section 3.1) are investigated and resolved through a series of steps described in Appendix C. Improvements are documented in technical reports along with expected impacts on the accuracy and cost of energy analysis methods. Organizations that maintain field data collection procedures and simulation protocols can choose whether to adopt improvements, depending on their specific accuracy requirements and cost limitations.

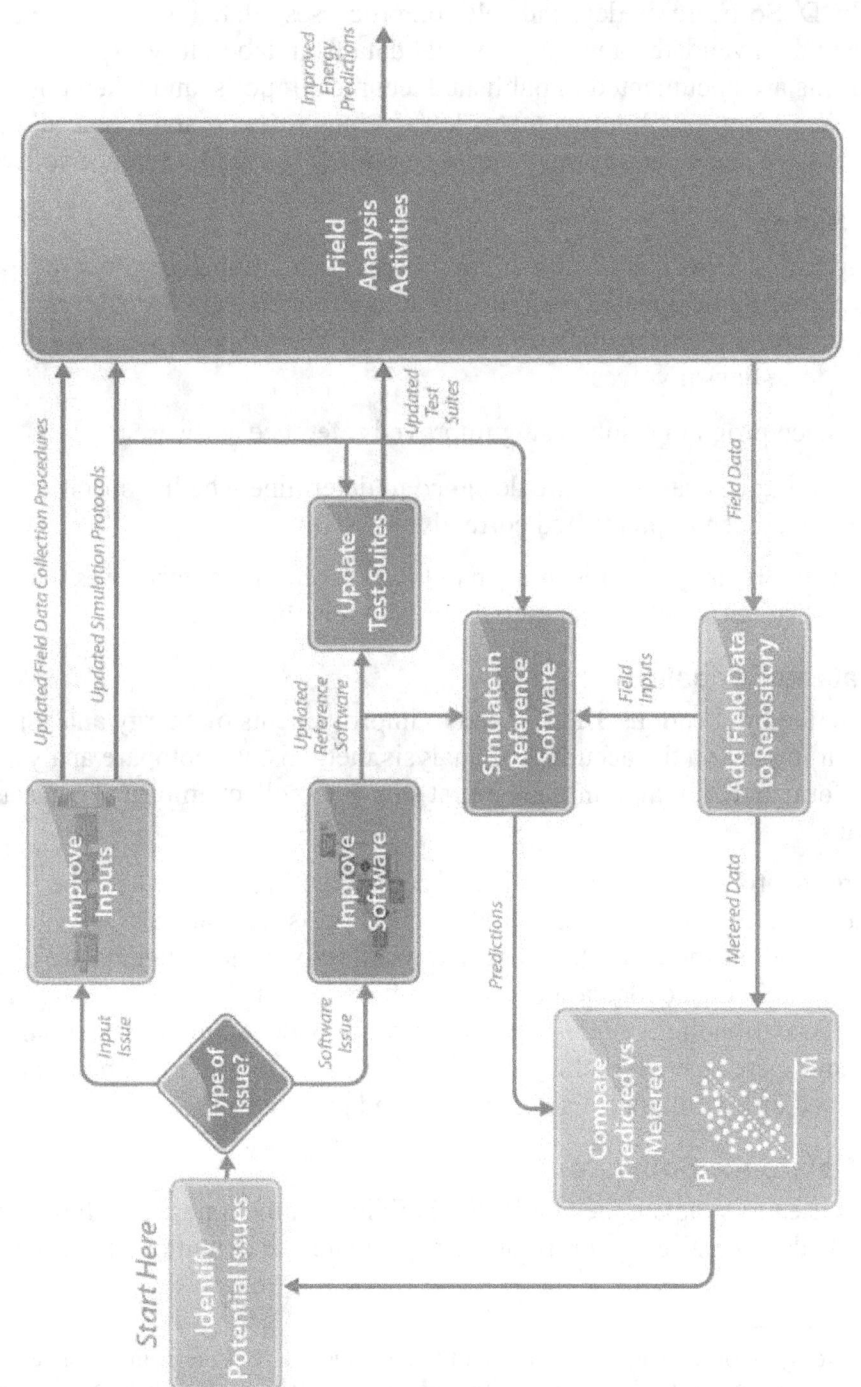

**Figure 2. Continuous process for improving the accuracy of energy analysis for residential buildings**

5

## 3.3 Improve Software

Potential software issues (see Section 3.1) are investigated and resolved through a series of steps described in Appendix D. Software models and solution processes related to the issue are isolated in the code and then validated against empirical data from laboratory experiments and field tests. Improvements are documented in published technical reports, and when possible, implemented in reference software.[2] Improvements to reference software tend to be additions or alterations to the underlying source code and result in compiling updated reference software.

## 3.4 Update Test Suites

Figure 2 shows the overall process that results in improvements to reference software. Comparative test suites should be updated periodically so commercial software developers can verify they have implemented software improvements and simulation protocols correctly.[3] Updating comparative test suites involves:

- Updating reference program results using improved reference software.

- Modifying or adding test cases that are designed to determine whether specific improvements have been implemented correctly.

Software improvement efforts may also result in analytical and empirical test cases that can be included in software test suites to help developers improve their software.

## 3.5 Add Field Data to Repository

The subprocesses described in Sections 3.1–3.4 lead to improvements of energy analysis methods. To track their impact on the accuracy of analysis methods, we compare analysis predictions to the performance of real homes. The first step is to collect empirical datasets from field analysis activities.

Field data for many residential buildings are collected and stored in a repository. The information collected in the field includes the building characteristics that are relevant to energy analysis, as well as metered utility data. The range and resolution of the field data are determined according to established field data collection procedures from labeling, rating, scoring, and retrofit efforts. The information in the repository must cover a wide range of residential building types and occupants and span each DOE climate region. When energy improvements are made, building descriptions and utility bill data also cover pre- and post-retrofit conditions.

## 3.6 Simulate in Reference Software

Reference software is used to simulate the residential buildings in the repository. Inputs come from the collected field data, or where appropriate, are calculated or defaulted under the

---

[2] NREL researchers are directly involved in the development of EnergyPlus, which is the primary reference software tool supported by DOE. Improvements to other tools will be made at the discretion of those developing and maintaining the software. As described in Section D.6, recommendations for improved models and the expected benefits of adopting improvements will be documented and published in technical reports to aid software developers. When selecting reference tools for comparative test suites, NREL will consider tools that are actively implementing bug fixes and model improvements.

[3] Comparative test suites cannot be used to verify that recommended field data collection procedures are being followed.

guidance of published simulation protocols. The simulations predict long-term energy use and energy savings for buildings and retrofit packages represented in the field data.

## 3.7 Compare Predicted Versus Metered

Reference software predictions of long-term energy use and energy savings are compared to the metered utility data in the repository. Utility bill analysis methods such as PRISM (Fels 1986) can be used to disaggregate the metered energy use into several estimated end-use bins, which provides more meaningful and diagnostic comparisons between predicted and metered data.[4] The aggregate accuracy of the energy analysis method is quantified as described in Appendix B. Because the current versions of the reference software are used in conjunction with current simulation protocols and field data collection procedures, the predictions represent the current state of residential energy analysis methods. If current energy analysis methods are not accurate enough, other potential sources of inaccuracy must be investigated.

---

[4] Direct comparison of predicted and measured energy use is not without challenges. Uncertainties associated with occupant behavior, weather normalization, and end-use disaggregation make direct comparison inexact. Reddy (2006) provides a good overview of these issues in the context of software calibration. Mills (2004) highlights limitations related to the availability and manipulation of measured data.

# 4   Ongoing Efforts

The purpose of this section is to describe ongoing research efforts at NREL and how each effort supports the method presented in Section 3.

## 4.1   Addressing Potential Issues

NREL has begun to identify, investigate, and correct input and software issues (see Figure 2). A preliminary list of potential issues was compiled and is presented in Appendix E, with short descriptions of each. NREL is also engaging the Building America Analysis Methods and Tools Standing Technical Committee to help:

- Identify potential issues.

- Prioritize potential issues.

- Peer review proposed improvements.

A series of short NREL technical reports will document results and recommendations for individual issues. Others will summarize improvements to field data collection procedures, simulation protocols, and software test suites.

## 4.2   National Residential Efficiency Measures Database

In 2009, DOE tasked NREL to develop a National Residential Efficiency Measures Database[5] to create standard technical definitions for energy retrofit measures so software developers and analysts would have access to consistent and vetted input information. The database is publicly available through a Web interface and XML (extensible markup language) feeds. Improvements to measure definitions will be made through scheduled database updates.

To further improve the database, NREL is collecting and organizing field data from residential retrofit programs. These characteristics, consumption, cost, and performance data will help evolve the public-facing, aggregate measure data. As described in Sections 3.5–3.7, NREL will use the field data repository to assess the accuracy of reference software under the current field data collection procedures and simulation protocols. NREL is striving to make field data available to commercial software developers so they can use large empirical datasets to track the impact of improvements on software accuracy.

## 4.3   Building America House Simulation Protocols

NREL will continue to maintain and improve the *Building America House Simulation Protocols*[6] (Hendron and Engebrecht 2010). Improvements will be documented in published updates, and can be adopted by other organizations that maintain simulation protocols.

## 4.4   BEopt™ Diagnostic Test Suite

To facilitate rapid comparison of research-level, whole-building simulation engines, NREL developed the BEopt Diagnostic Test Suite. BEopt is a building energy optimization tool that currently interfaces with two simulation engines: EnergyPlus and DOE-2.2. Because BEopt is designed to evaluate alternative energy efficiency options and retrofit measures in new home construction and existing buildings, respectively, a comprehensive range of building

---

[5] See http://www.nrel.gov/ap/retrofits/index.cfm.
[6] Previously called the *Building America Benchmark*.

characteristics, site conditions, and occupant behavior can be simulated automatically and systematically. This BEopt Diagnostic Test Suite allows researchers at NREL to identify and understand differences between simulation engines by eliminating discrepancies caused by inputs. Discrepancies between simulation engines may indicate potential software issues where analytical verification and empirical validation are needed.[7]

## 4.5 Building Energy Simulation Test

NREL researchers have developed an overall building energy software validation methodology that includes analytical solutions, comparative testing, and empirical data (Judkoff and Neymark 2006). The methodology has been adopted by the American Society of Heating, Refrigerating and Air-Conditioning Engineers (ASHRAE) Standard 140 and is discussed in the ASHRAE *Handbook of Fundamentals*, along with the pros and cons of the analytical, comparative, and empirical methods (ASHRAE 2005). NREL researchers have written several building energy simulation test (BESTEST) suites. Most have used comparative methods where software predictions are evaluated versus reference software predictions, and versus analytical solutions and verified numerical solutions when possible (see Section D.3). These test suites have helped software developers diagnose and fix discrepancies caused by physical algorithms and coding. However, because there is no truth standard in comparative testing, such a suite is only as good as the reference software that is the basis for comparison. Also, analytical solutions and verified numerical solutions represent a mathematical truth standard, but not necessarily a physical truth standard. NREL is therefore interested in addressing empirical validation, the third component of the overall validation method.

NREL will develop test suites that can be used to directly compare software predictions to empirical data. Classes of empirical data include:

- **Laboratory and field test data for isolated building subsystems.** Empirical datasets collected to validate software algorithms (see Appendix D) may be presented in a standardized form as empirical test cases. Such data, which are collected from highly controlled experiments designed to minimize input error, can be used to validate algorithms corresponding to different subsystems and the interactions of subsystems in building energy simulation programs.

- **Building characteristics and utility bill data.** Empirical datasets collected to support the National Residential Efficiency Measures Database may also be presented in a standardized form to aid software developers in the evaluation of their tools against measured data. This type of data, which is collected from field analysis activities where input errors are not tightly controlled, can be used to assess the performance of software as it is typically used in the field.

The purpose of developing software test suites is to provide tools for developers to improve the accuracy of their software. Therefore, test suites should be diagnostic (allow users to determine the sources of inaccuracy) and cover a range of building characteristics, site conditions, and

---

[7] Agreement between two simulation engines does not ensure agreement with the true performance of buildings; some models and solution processes in both programs may be incorrect relative to the true physical behavior of the building system. Possible software issues are thus identified by applying other techniques in addition to comparative testing.

occupant behavior. NREL will periodically update BESTEST comparative suites to reflect improvements to simulation protocols and reference software (see Section 3.4).

## 4.6   Estimating Uncertainty in Energy Analysis Predictions

Whether an input is defaulted, estimated, or measured, there is some uncertainty as to how closely the input value matches the "true" value. These errors in input propagate through a building energy simulation program to produce errors in output. Because there will always be some uncertainty in input values, there will always be some uncertainty in software predictions.

In the process of addressing specific input issues, NREL will estimate the uncertainty of many input values for software (see Appendix C). NREL is also exploring methods to estimate the uncertainty and natural variability in energy assessment data through controlled field studies. As more information is collected, NREL will use batch Monte Carlo simulation methods to estimate the overall uncertainty in predicted energy uses and savings for a variety of conditions. Understanding the variability and uncertainty in predictions will help establish expectations for software accuracy and identify the relative importance of each input parameter. Results from this research will lead to more accurate, streamlined, and cost-effective energy assessments.

## 4.7   Field Testing

NREL and Building America field tests will collect data that can be used to validate software models (Section 3.3) and improve input collection methods (Section 3.2).

## 4.8   Thermal Test Facility Experiments

Measurements taken in the NREL Thermal Test Facility will be used to improve the accuracy of models and inputs for various heating, ventilation, and air-conditioning and domestic hot water systems. For example, laboratory tests will yield performance maps for various new and existing air-conditioning systems, which will help us validate software models and improve default input parameters. These detailed performance characteristics will be added to the National Residential Efficiency Measures Database as they become available.

## 4.9   Field Data Measurement Technology Research and Development

In the process improving field data collection procedures (Section 3.2), gaps may be identified that demonstrate the need for measurement technology research and development. NREL plans to develop and test alternative measurement technologies that will lead to more accurate, more affordable, faster, and safer energy assessments.

# 5 Final Remarks

A method for improving the accuracy of residential energy analysis methods has been presented. Application of the method will result in improvements to:

- Field data collection procedures
- Simulation protocols
- Software test suites.

Energy assessors, analysts, and software developers can incorporate these improvements into their activities, resulting in more accurate energy analysis in the field. More accurate energy analysis methods will improve the success of energy-efficient design, labeling, scoring, rating, and retrofit efforts.

# 6 References

ASHRAE (2001). *ANSI/ASHRAE Standard 140-2001.* "Standard Method of Test for the Evaluation of Building Energy Analysis Computer Simulation Programs." Atlanta, GA: American Society of Heating, Refrigerating and Air-Conditioning Engineers.

ASHRAE (2004). *ANSI/ASHRAE Standard 140-2004.* "Standard Method of Test for the Evaluation of Building Energy Analysis Computer Simulation Programs." Atlanta, GA: American Society of Heating, Refrigerating and Air-Conditioning Engineers.

ASHRAE (2005). *ASHRAE Handbook Fundamentals.* Atlanta, GA: American Society of Heating, Refrigerating and Air-Conditioning Engineers.

ASHRAE (2007). *ANSI/ASHRAE Standard 140-2007.* "Standard Method of Test for the Evaluation of Building Energy Analysis Computer Simulation Programs." Atlanta, GA: American Society of Heating, Refrigerating and Air-Conditioning Engineers.

Berry, L.; Gettings, M. (1998). "Realization Rates of the National Energy Audit." In *Proceedings of Thermal Performance of the Exterior Envelopes of Buildings VII.* Clearwater, Florida: American Society of Heating, Refrigerating and Air-Conditioning Engineers.

Earth Advantage Institute, Conservation Services Group (EAI/CSG). (2009). *Energy Performance Score (EPS) 2008 Pilot Findings & Recommendations Report.* Energy Trust of Oregon.

Fels, M.F. (1986). "PRISM: An Introduction." *Energy and Buildings* 9:5–18.

Francisco, P.; Palmiter, L. (1996). "Modeled and Measured Infiltration in Ten Single-Family Homes." ACEEE 1996 Summer Study of Energy Efficiency in Buildings. Washington, DC: American Council for an Energy Efficient Economy.

Hendron, R.; Engebrecht, C. (2010). *Building America House Simulation Protocols.* Golden, CO: National Renewable Energy Laboratory. NREL/TP-550-49246.

Judkoff, R.; Neymark, J. (1995a). *International Energy Agency Building Energy Simulation Test (BESTEST) and Diagnostic Method.* Golden, CO: National Renewable Energy Laboratory. NREL/TP-472-6231.

Judkoff, R.; Neymark, J. (1995b). *Home Energy Rating System Building Energy Simulation Test (HERS BESTEST). Vol. 1: Tier 1 and Tier 2 Tests User's Manual and Vol. 2: Tier 1 and Tier 2 Tests Reference Results.* Golden, CO: National Renewable Energy Laboratory. NREL/TP-472-7332.

Judkoff, R.; Neymark, J. (2006). "Model Validation and Testing: The Methodological Foundation of ASHRAE Standard 140." *ASHRAE Transactions* 112(2):367–376. Atlanta, GA: American Society of Heating, Refrigerating and Air-Conditioning Engineers.

Judkoff, R.; Polly, B.; Bianchi, M.; Neymark, J. (2010). *Building Energy Simulation Test for Existing Homes (BESTEST-EX), Phase 1 Test Procedure: Building Thermal Fabric Cases.* Golden, CO: National Renewable Energy Laboratory. NREL/TP-550-47427.

Judkoff, R.; Wortman, D.; O'Doherty, B.; Burch, J. (2008). *A Methodology for Validating Building Energy Analysis Simulations.* Golden, CO: National Renewable Energy Laboratory.

NREL/TP-550-42059. Based on unpublished report of 1983 with same authors and title, previously referenced as SERI/TR-254-1508.

Mills, E. (2004). "Inter-comparison of North American Residential Energy Analysis Tools." *Energy and Buildings* 36:865–880.

Neymark, J.; Girault, P.; Guyon, G.; Judkoff, R.; LeBerre, R.; Ojalvo, J.; Reimer, P. (2005). "The 'ETNA BESTEST' Empirical Validation Dataset". Building Simulation 2005, Ninth International IBPSA Conference, 15–18 August, 2005; Montreal, Canada. International Building Performance Simulation Association.

Neymark, J.; Judkoff, R. (2002). *International Energy Agency Building Energy Simulation Test and Diagnostic Method for Heating, Ventilating, and Air Conditioning Equipment Models (HVAC BESTEST), Volume 1: Cases E100-E200.* Golden, CO: National Renewable Energy Laboratory, NREL/TP-550-30152.

Neymark, J.; Judkoff, R. (2004). *International Energy Agency Building Energy Simulation Test and Diagnostic Method for Heating, Ventilating, and Air Conditioning Equipment Models (HVAC BESTEST), Volume 2: Cases E300-E545, E200.* Golden, CO: National Renewable Energy Laboratory, NREL/TP-550-36754.

Neymark, J.; Judkoff, R. (2008). *International Energy Agency Building Energy Simulation Test and Diagnostic Method (IEA BESTEST), Multi-Zone Non-Airflow In-Depth Diagnostic Cases: MZ320 -- MZ360.* Golden, CO: National Renewable Energy Laboratory, NREL/TP-550-43827.

Neymark, J.; Judkoff, R. (2009a). *International Energy Agency Building Energy Simulation Test and Diagnostic Method (IEA BESTEST) In-Depth Diagnostic Cases for Ground Coupled Heat Transfer Related to Slab-On-Grade Construction.* Golden, CO: National Renewable Energy Laboratory, NREL/TP-550-43388.

Neymark, J.; Judkoff, R. (2009b). "IEA BESTEST In-Depth Diagnostic Cases for Ground Coupled Heat Transfer Related to Slab-On-Grade Construction." *Building Simulation 2009*, 27-30 July, 2009; Glasgow, Scotland , UK. International Building Performance Simulation Association. Golden, CO: National Renewable Energy Laboratory, NREL/CP-550-45742.

Pigg, S.; Nevius, M. (2000). *Energy and Housing in Wisconsin: A Study of Single-Family Owner-Occupied Homes.* Energy Center of Wisconsin, 199-1.

Reddy, T.A. (2006). "Literature Review on Calibration of Building Energy Simulation Programs: Uses, Problems, Procedures, Uncertainty, and Tools" *ASHRAE Transactions* 112(1):226-240. Atlanta, GA: American Society of Heating, Refrigerating and Air-Conditioning Engineers.

Roberts, D.R.; Blasnik, M. (2005). *Real vs. Rated Energy Use.* Roundtable held at 2005 Affordable Comfort Conference, Indianapolis, IN.

Stein, J.R. (1997). *Accuracy of Home Energy Rating Systems.* LBNL/40394.

Sterling, R.L.; Gupta, S.; Shen, L.S.; Goldberg, L.F. (1993). *Assessment of Soil Thermal Conductivity for Use in Building Design and Analysis.* Report to the American Society of Heating, Refrigerating and Air-Conditioning Engineers. Atlanta, GA.

Tabares-Velasco, P.C. and Griffith, B. (2011). "Diagnostic Test Cases for Verifying Surface Heat Transfer Algorithms and Boundary Conditions in Building Energy Simulation Programs." Accepted in *Journal of Building Performance Simulation*.

Ternes, M.P. (2007). Validation *of the Manufactured Home Energy Audit (MHEA)*. ORNL/CON-501.

Ternes, M.P; Gettings, M.B. (2008). *Analyses to Verify and Improve the Accuracy of the Manufactured Home Energy Audit*. ORNL/CON-506.

# Appendix A    Literature Review

Anecdotal evidence from the field and controlled studies have raised concerns about the accuracy of software-based energy analysis for existing homes. Previous studies showing evidence of overprediction are discussed in Section A.1 and possible causes for overprediction are discussed in Section A.2. Conclusions regarding both sections are presented in Section A.3.

## A.1  Previous Studies Showing Evidence of Overprediction

Overprediction of energy use and savings by residential energy analysis methods has been observed in previous studies. A summary of some examples in the literature occurring before 1999 can be found in Berry and Gettings (1998). Other examples and more recent studies are discussed below.

A validation study performed at the Energy Center of Wisconsin compared predicted results from Home Energy Rating System (HERS) software to space heating bills for 147 homes in Wisconsin (Pigg and Nevius 2000). The homes ranged from 1930s pre-retrofit to 1990s ENERGY STAR® qualified. The study clearly indicated that the simulation software overpredicted space heating energy use in homes with higher space heating bills, likely older, poorly insulated, leaky homes (see Figure 3).

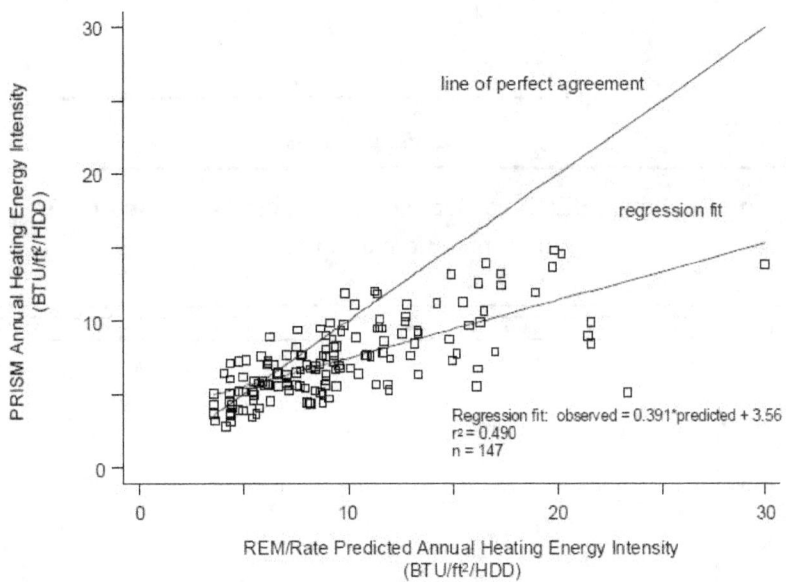

**Figure 3. Overprediction of space heating energy use observed in Energy Center of Wisconsin Study**

(reprinted with permission from Pigg and Nevius 2000)

The Energy Trust of Oregon performed a study to evaluate building energy simulation programs. Three programs were compared: SIMPLE, REM/Rate, and Home Energy Saver (HES) (EAI/CSG 2009). Detailed audits were conducted and utility bills were collected for 190 homes in Portland and Bend. The homes were simulated with the three simulation tools, including two levels of detail for HES. On average REM/Rate (SERI/RES-based) and HES (DOE-2.1E-based) predicted energy use higher than the utility bills. In particular, the simulation programs overpredicted, to a greater degree, energy use in older homes, which tend to be poorly insulated and leaky. For example, the detailed HES audit and simulation overpredicted gas use for space heating by an average of 41% in homes built before 1960, versus 13% overprediction for homes built after 1989.

Stein (1997) conducted a study to examine the accuracy of HERS software in predicting energy use in homes. As part of the study, he looked closely at 200 homes receiving HERS ratings under the California HERS. Stein compared predicted energy use to utility bill data for the 200 homes. He concluded that on average the simulations overpredicted energy use in the homes, and the overprediction increased with house age (see Figure 4 and Figure 5).

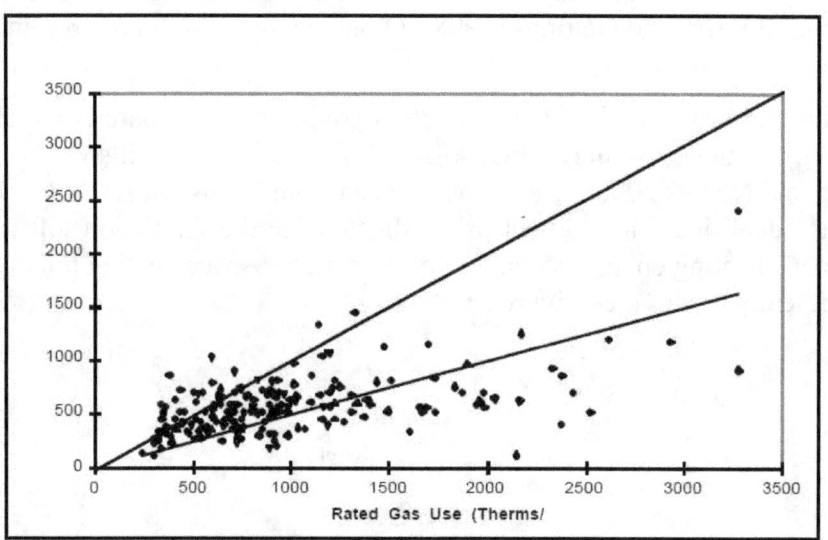

**Figure 4. Stein found California HERS software overpredicted gas use (y-axis is actual gas use)**
(reprinted from Stein 1997)

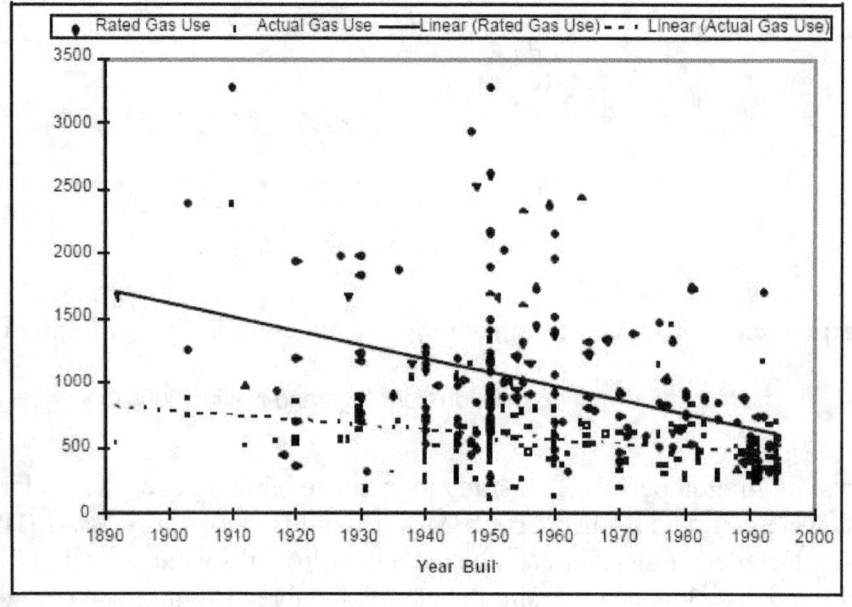

**Figure 5. Stein found that overprediction increased with house age**
(reprinted from Stein 1997)

16

A two-part study by Oak Ridge National Laboratory (ORNL) documents validation of the Manufactured Home Energy Audit tool (Ternes 2007; Ternes and Gettings 2008). A representative set of audit and utility data on manufactured homes across five northern/mid-western states was selected for analysis. Predicted space heating energy use was compared with actual space heating energy use extracted from billing data collected for the study. Comparisons between modeled and actual energy use were made pre- and post-retrofit. Pre-retrofit simulation results were found to overpredict space heating. Post-retrofit modeled results were within an acceptable range. As a result, energy savings resulting from the retrofit were shown to be overpredicted.

## A.2  Possible Causes of Overprediction

When predicted whole-building energy use and savings are compared to actual energy use and savings, it is not always clear whether inaccuracies are attributable to (Berry and Gettings [1998] and Judkoff and Neymark [2006]):

- Misuse and misunderstandings of energy analysis methods
- Deficiencies in inputs
- Deficiencies in software
- Causes external to the analysis method.

### A.2.1  Misuse and Misunderstandings

In some cases, especially those that result in anecdotal evidence, analysis methods are possibly being misused or misunderstood. For example, an *asset evaluation* involves modeling the energy use of a house under standardized occupant assumptions (so results from one house can be compared fairly to another), whereas an *operational evaluation* considers the specific occupant behavior. It would be a misuse and misunderstanding of the analysis method to perform an asset evaluation on a single home, provide recommendations for efficiency improvements to the specific occupants, and then expect close agreement between predicted and measured energy savings for that home. Savings can be significantly changed by occupant behavior.

Misunderstandings may also contribute to perceived inaccuracy. Although *accuracy* is often used as a general term that describes the agreement between predicted and metered energy usage, the perception of accuracy can vary greatly across the parties involved in the analysis. For example, an analysis method may perform accurately on average across a large population of homes in a typical year, but perform inaccurately for individual homes or in a year with atypical weather. In this situation, individual homeowners may be unsatisfied when the analysis method produces inaccurate results for their homes, even though a program manager may be satisfied with the method's ability to accurately predict overall energy savings for a program.

### A.2.2  Input Deficiencies

Inaccuracies in defaulted, estimated, or measured inputs may lead to disagreement between predicted and actual energy use and savings. When homes are modeled based on energy assessment information alone, there is little control of input error (compared to detailed laboratory field testing where advanced measurement techniques are used to reduce input errors). Discrepancies could result from inaccurate measurements and observations, inaccurate handbook values for material properties, inaccurate default input values in software (either user-defined or hard-wired), errors translating input information into the building simulation model, etc.

### A.2.3  Software Deficiencies

Inaccuracies in energy model predictions may also be due to software deficiencies. The software may do a poor job at modeling the physical behavior of the building because of inappropriate models or software coding errors. Because simulation programs can have hundreds of thousands of lines of code, and because inputs are not fully controlled in the studies described above, identifying or attributing specific inaccuracies in models is difficult. Statistical analysis across various building types, occupants, and locations may help identify sources of error, but a more detailed and controlled approach is needed to isolate the error to specific algorithms within the code.

### A.2.4  Causes External to the Analysis Method

Discrepancies may also be caused by inaccuracies and inconsistencies external to the analysis method. For example, retrofit measures may not be installed correctly or consistently, as modeled in the software. Furthermore, there may be inaccuracies in the processing of metered data (disaggregation, weather normalization, etc.), against which software predictions are compared.

## A.3  Conclusions

The following conclusions and observations can be made based on the literature review:

- Multiple studies confirm that analysis methods tend to overpredict energy use and savings in poorly insulated, leaky homes with older mechanical systems, that is, homes most needing energy retrofits.

- Determining the reason for discrepancies is difficult, because all error sources act simultaneously.

- Many metrics for accuracy are used in various studies and many possible causes of overprediction are described. A single reference containing definitions for error, potential sources of error, and accuracy, developed in the context of comparing simulated building energy use to measured energy use, would be beneficial.

- A method to improve the accuracy of energy use and savings predictions is needed.

# Appendix B Definitions and Calculation Methods for Error and Accuracy

The purpose of this Appendix is to define error, identify the possible sources of error, and discuss some metrics that can be used as indicators of accuracy. These definitions are developed in the context of comparing predicted energy use to measured energy use in residential buildings.

## B.1 Definition of Error

For the purpose of building energy analysis, error ($e$) can be defined as the difference between a predicted value ($p$) and a measured value ($m$), such that

$$e = p - m \qquad \qquad \text{(B1)}$$

Error can be calculated for a variety of predicted and measured values, including energy uses, temperatures, and heat flows.

## B.2 Possible Sources of Error

Possible sources of error can be grouped in to the following categories, which are based on Judkoff and Neymark (2006)[8] and Berry and Gettings (1998):

### B.2.1 Errors Related to Analysis Methods

**Input Errors:**

- **Building Inputs.** Inaccuracies in the description of building geometry, material physical properties, and characteristics of mechanical equipment.

- **Occupant Inputs.** Inaccuracies in the description of occupant behavior and occupant-controlled equipment settings.

- **Site Inputs.** Inaccuracies in the description of the local weather, soil, and adjacent structures/vegetation.

**Software Errors:**

- **Physics Algorithms.** Inaccuracies in the mathematical modeling of the physical behavior of the building and its equipment.[9]

- **Coding Errors.** Typographical and logic errors inadvertently introduced into the software code.

### B.2.3 Errors External to Analysis Methods

This report presents methods for reducing sources of error related to analysis methods. However, other research efforts are needed to:

---

[8] Judkoff and Neymark (2006) referred to input and software errors as "external" and "internal" errors, respectively.

[9] There are two types of physics errors: i) the basic mechanism is not well-understood (e.g. infiltration or moisture transport); and ii) implementation is simplified (e.g. ground coupling or interior infrared radiation exchange).

- Reduce inaccuracies in the process of analyzing metered data when comparing to long-term predictions of energy use or savings[10] (e.g. weather normalization, aggregation, disaggregation).

- Reduce inconsistencies in the construction and retrofit of homes by properly training construction, installation, and retrofit workers and establishing Best Practices Manuals.

## B.3   Calculating Errors for Analysis of a Single Home

It is easiest to understand and discuss error in the context of energy analysis for a single home. As an example, if an energy analysis method predicts 100 MMBtu/yr of energy use for an existing home, and the metered energy use is 80 MMBtu/yr, the error is +20 MMBtu/yr. Error can also be calculated more frequently based on predicted and measured values. For example, one could calculate the hourly error. In this case there would be 8,760 predicted values per year, each of which would be compared to a measured value, yielding 8,760 error values per year.

When multiple error values are calculated, visual inspection and statistical methods can be used to study the nature of the errors. Errors generally have two components:

- **Bias Error.** A systematic trend of disagreement between predicted and measured values, either too high or too low.

- **Random Error.** Variation in error that does not contribute to a trend of disagreement.

The sum of these two components gives the error ($e$). Multiple errors must be observed to estimate the bias and random components of each. Bias errors can be identified by looking for an average offset or trends in error as a function of another variable such as time or temperature. For example, Figure 6 indicates the presence of a bias error because time-averaged error is above zero and the individual errors follow a nonzero trend (the solid line). Random errors can be identified by looking at the variation of errors above and below the trends. In Figure 6, the random error is the fluctuation of the error about the offset trend line. Because the average value of random errors is zero (by definition), the magnitude of the random scatter is of primary interest.

---

[10] Some analysis methods (e.g., inverse models) may use metered data directly, or may calibrate inputs or adjust outputs based on metered data. In these cases, inaccuracies in metered data or the processing of those data are sources of error related to the analysis method.

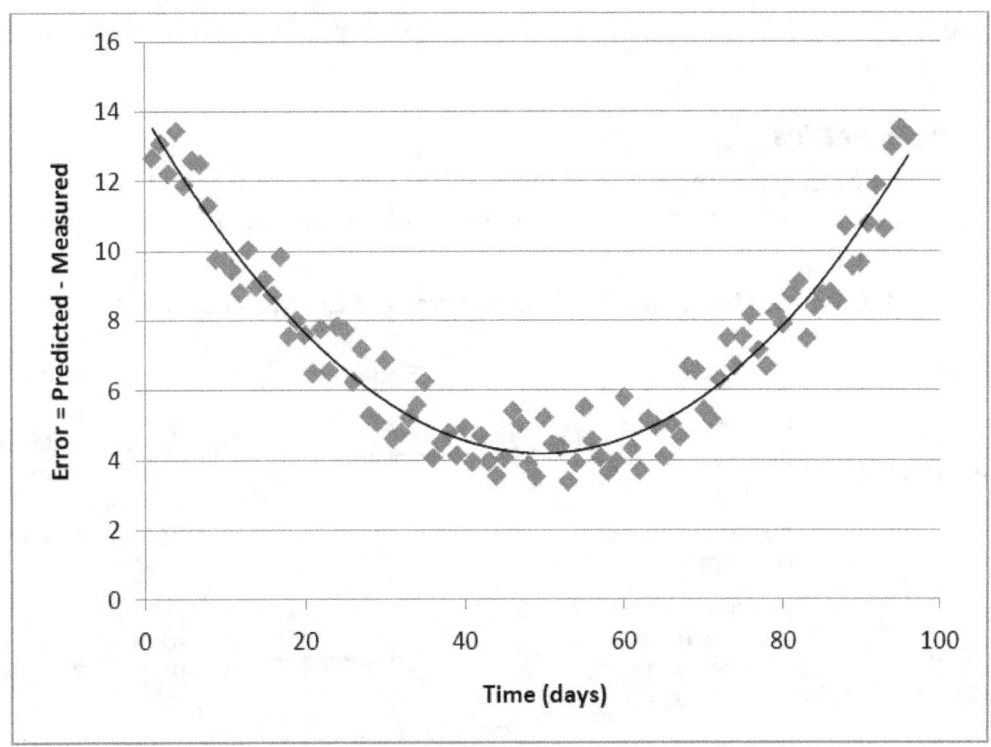

**Figure 6. Example plot of errors between predicted and measured values for a single home**

## B.4 Calculating Errors for Analysis of Multiple Homes

As seen in Appendix A, it is common to discuss the errors of predictions over multiple buildings, occupants, and sites.[11] For each home a predicted value (e.g., annual energy use) is compared to a measured value to calculate an error corresponding to that specific home. The calculation can be repeated for many homes and the individual errors can be compared.

Comparing across many homes has advantages:

- Bias and random error due to inputs become distinguishable as more homes are included in the comparison.

- Trends in errors can be identified as a function of input variables (such as foundation type, retrofit measure, climate, occupant type, etc.) to help identify potential sources of error.

However, error analysis for multiple homes usually involves a single prediction for each home (e.g., annual energy use), which is a function of most inputs and algorithms in the simulation program. This can make it difficult to isolate errors to specific inputs and software algorithms, especially if the number of homes is not sufficient to conduct meaningful statistical analysis.

## B.5 Metrics for Accuracy

With a clear definition of error, one can define metrics that track the accuracy of energy analysis methods. Accuracy metrics are generally calculated based on errors between predicted and

---

[11] We refer to a unique set of building, occupant, and site inputs as a "home" in this discussion, because typically software predictions are compared to measured data across a number of homes.

measured values. Like errors, accuracy metrics can be calculated for a single home or across many homes.

## B.5.1 Example Metrics

Examples of accuracy metrics are presented in Table 1 given a set of $n$ calculated errors. This example set is an initial list[12]; other metrics may be useful for tracking the accuracy of software predictions.

**Table 1. Example Metrics for Tracking Accuracy of Energy Analysis Methods**

| Metric | Description | Equation | Notes |
|---|---|---|---|
| Minimum and maximum errors | The minimum and maximum values of the errors | Self explanatory | Indicate extreme values of error |
| Mean error | The mean value of the errors | $\dfrac{\sum_{i=1}^{n} e_i}{n}$ | Primarily tracks overall bias error |
| Mean percent error* | The mean value of the errors expressed as percentages of measured values ($m_i$) | $\dfrac{\sum_{i=1}^{n}(e_i/m_i) \times 100}{n}$ | Primarily tracks overall bias error |
| Median error | The median value of the errors | The value for which 50% of errors are lower and 50% are higher. | Primarily tracks overall bias error |
| Nonzero mean error | The mean value of nonzero errors | $\dfrac{\sum_{i=1}^{k} e_i}{k}$ (k = number of nonzero errors) | Primarily tracks overall bias error without being affected by zero values |
| Standard deviation of the errors | The sample standard deviation of the errors | $\sqrt{\dfrac{1}{n-1}\sum_{i=1}^{n}(e_i - \bar{e})^2}$ $\bar{e} = mean\ error$ | Primarily tracks magnitude of random error |
| Mean absolute error | The mean value of the absolute errors | $\dfrac{\sum_{i=1}^{n}|e_i|}{n}$ | Does not distinguish between over- and underprediction |
| Median absolute error | The median value of the absolute errors | Self explanatory | Does not distinguish between over- and underprediction |
| Root mean squared error (RMSE) | The square root of the mean value of the squared errors | $RMSE = \sqrt{\dfrac{\sum_{i=1}^{n} e_i^2}{n}}$ | Does not distinguish between over- and underprediction |
| Normalized root mean squared error (NRMSE) | Normalized square root of the mean value of the squared errors | $NRMSE = \dfrac{RMSE}{Range\ or\ Mean}$ | Does not distinguish between over- and underprediction |

\* Other metrics in this table can be calculated based on percent errors, but are not listed here to limit repetition.

---

[12] EIA/CSG (2009) is an example of a study that reports some of the metrics listed in Table 1.

## B.5.2 Selecting a Metric To Track Accuracy of Analysis Methods

Accuracy metrics should be selected based on the purpose for evaluating the performance of the energy analysis method(s). For example, if a retrofit program manager is concerned about the accuracy of individual predictions (to ensure each participant achieves predicted savings), as well as the average accuracy of many predictions (to ensure program goals are met), a metric or a combination of metrics should be selected that captures bias and random errors.

# Appendix C   "Improve Inputs" Subprocess Description

In previous NREL validation efforts, inputs were highly defined and controlled to investigate the internal workings of software. However, the accuracy of residential audit tools, as they are applied in the field, also depends on the accuracy of methods for defaulting, measuring, and estimating simulation inputs for residential buildings.[13] Hundreds of inputs go into a building simulation, but their relative importance and required accuracy have not been widely established.

Figure 7 illustrates a *continuous input improvement method for whole-building residential energy analysis*. Each segment of the illustrated method is explained in the following subsections.

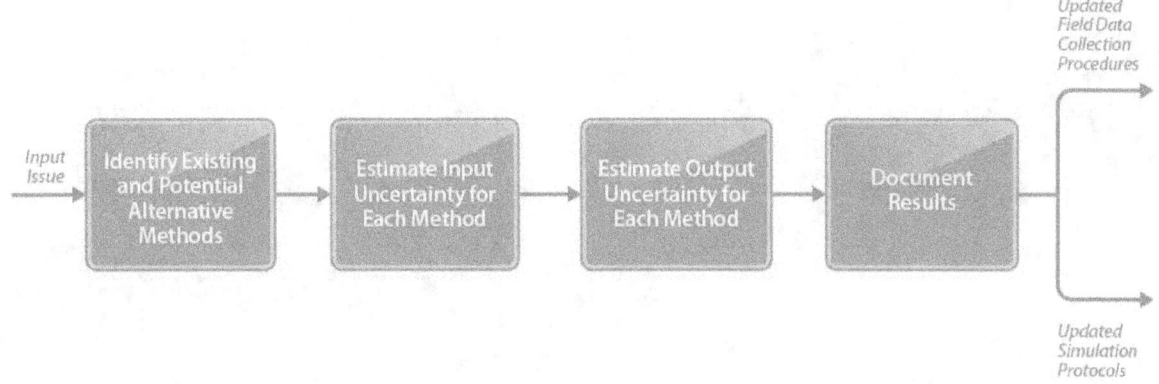

**Figure 7. Continuous input improvement method for whole-building residential energy analysis**

## C.1   Identify Existing and Potential Alternative Methods

The first step in resolving a potential input issue is to clearly identify which software inputs are directly relevant to the issue. For example, if the potential issue is "Thermostat Setback," the key software inputs are the thermostat settings for the heating and cooling seasons. Once the inputs have been identified, existing methods for defaulting, estimating, and measuring the inputs are surveyed. Alternative input collection methods may also be hypothesized and investigated. The goal is to develop a list of potential input collection methods that includes existing methods and covers a range of approaches, from defaulting, to estimating based on other observations or measurements, to directly measuring the input values.

## C.2   Estimate Input Uncertainty for Each Method

Once a list of potential input collection methods is identified, the next step is to estimate the uncertainty in the collected inputs for each method. Controlled studies or previous datasets may be needed to estimate the uncertainty in inputs. If the input is measured directly, the uncertainty can be estimated based on the measurement uncertainty, and the uncertainty that the measured value can be extrapolated over the analysis period (e.g. if set points are measured over one week periods during the heating and cooling seasons, there is uncertainty in the extrapolation of measured values throughout the heating and cooling seasons).

---

[13] Previous efforts related to the accuracy of inputs primarily fell under the development of the *Building America House Simulation Protocols* (Hendron and Engebrecht 2010). Section 3 outlines a plan to closely link software validation efforts with the development of simulation protocols.

The goal is to estimate statistical distributions of input uncertainties for each method. Continuing with the example above, one method for collecting heating and cooling set points may be defaulting to 68°F for the heating set point and 78°F for cooling set point. A large dataset with measured thermostat settings for a variety of homes could be analyzed to estimate the statistical distribution of errors when using these simple default values in place of the true values. If the errors follow a normal distribution, the uncertainty could be characterized by a standard deviation (e.g., ± 4°F). The uncertainty may not follow a normal distribution, so more advanced statistical methods may be needed.

## C.3   Estimate Output Uncertainty for Each Method

For each input collection method, the uncertainties in inputs can be propagated through the software to estimate the resulting uncertainties in predictions of energy use and savings.[14] The uncertainties in predictions can change depending on the values of other inputs (e.g., climate, envelope performance, and system efficiency), so the error propagation analysis should cover a range of building types, occupant behavior, and site conditions. In initial analyses, uncertainties for all simulation inputs will not propagated simultaneously, so the uncertainty in output will be a minimum estimate for the overall output uncertainty. As more information about input uncertainties is gathered through the continuous improvement process, more comprehensive error propagation studies can be conducted to better understand the contributions of specific input uncertainties to the overall output uncertainty.

In some cases, it may be possible to use empirical datasets to directly estimate the output uncertainty for different input collection methods. If an empirical dataset contains input information from energy assessments and utility billing data, the benefits and tradeoffs of some input collection methods can be investigated by simulating the buildings and analyzing errors (similar to comparisons described in Section 3.7). For example, if the dataset contains $CFM_{50}$ blower door measurements, the accuracy of the analysis method using that input collection method can be evaluated and compared to other methods of defaulting and estimating the required infiltration inputs. In these cases the errors and uncertainties estimated using empirical data will include all other sources of error. Additionally, one cannot test input collection methods that require measurements or observations that were not made during the original energy assessment without returning to the homes to collect the relevant information.[15]

## C.4   Document Results

The primary purpose of the input improvement process to document tradeoffs between different input collection methods. Specific topics include:

- The impact on output uncertainty when defaulted versus measured versus estimated

- The time required to make estimations or take measurements

- The time and money required to train people to make estimations or take measurements

- The cost and availability of measurement equipment.

---

[14] Monte Carlo simulation techniques can be used to study the propagation of errors for a variety of building types, occupant behavior, and site conditions.

[15] It would be important to confirm that inputs collected at a later time are representative of the inputs corresponding to the time during which utility billing data were collected.

The "best" input collection method depends largely on the purpose and constraints of the analysis. For example, if the purpose is to rate the energy performance of a home on a 1–4 scale, the allowable output uncertainty may be higher than for an analysis used to predict retrofit savings for an improvement financed by the homeowner. The allowable cost of field data collection procedures can also vary depending on the application. Thus, data collection methods are documented; this informs the improvement of field data collection procedures and simulation protocols.

# Appendix D    "Improve Software" Subprocess Description

In the early 1980s (Judkoff et al. 2008), NREL developed a methodological foundation for evaluating the accuracy of whole-building energy simulation programs. The NREL methodology has been adopted by other organizations, including ANSI/ASHRAE Standard 140 (ANSI/ASHRAE 2007, 2004, 2001), *Standard Method of Test for the Evaluation of Building Energy Analysis Computer Programs*, which was, to our knowledge, the first codified method of test for building energy software in the world. Research efforts have resulted in numerous software test suites to aid software developers. The method presented in this section is an adaptation and application of the NREL validation methodology. Figure 8 illustrates a *continuous improvement method for residential whole-building energy simulation software*. Each segment of the method is discussed in the following subsections.

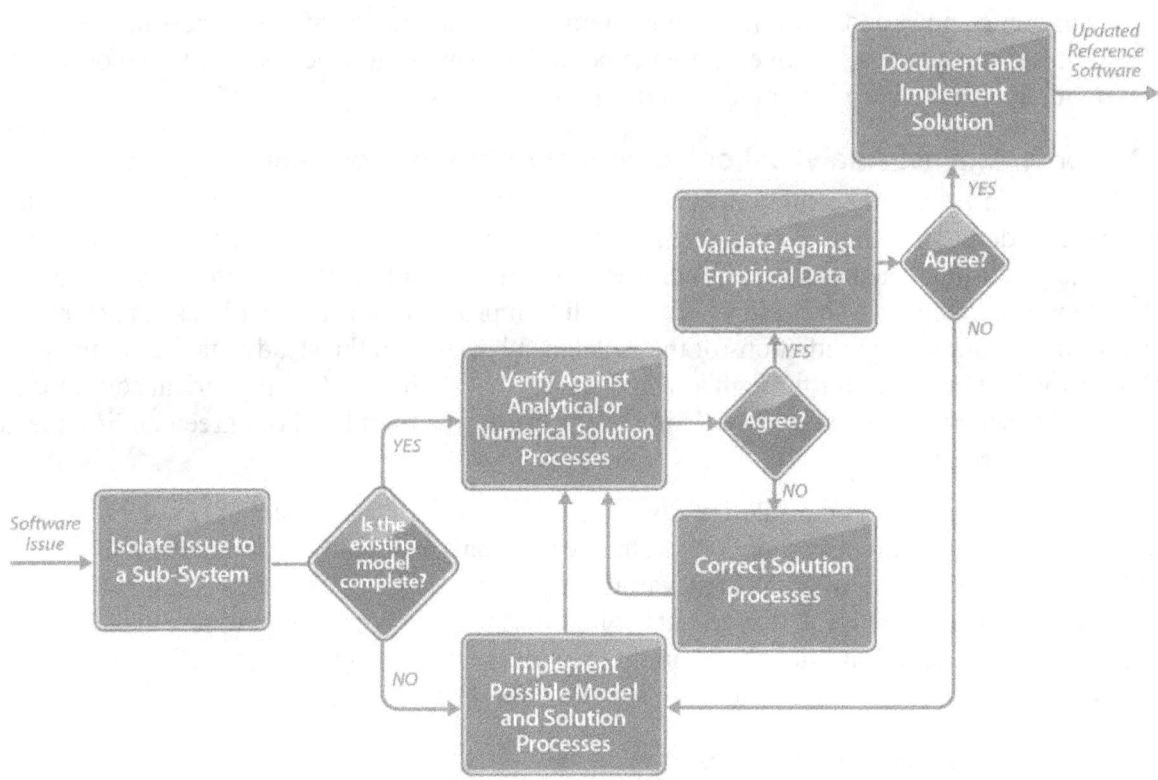

**Figure 8. Continuous improvement method for residential whole-building energy simulation software**

## D.1  Isolate Issue to a Subsystem

A building energy simulation program can have hundreds of thousands of lines of code corresponding to hundreds of models and solution processes for various building subsystems and the interactions of subsystems with each other, the occupants, and the building site. Subsystems contain a specific set of models and solution processes. Given a potential software issue, the first step is to isolate the issue to a specific subsystem to make it easier to verify model solution processes and validate models against measured data. Some considerations that are needed to define an isolated subsystem follow.

- All inputs for the model(s) in the subsystem must be identifiable and controllable.

- The model(s) in the subsystem must have an analytical solution or verified numerical solution under controlled input conditions.

- It must be possible to study the subsystem in a controlled experiment or field test with inputs to the subsystem either controlled or measured.

Isolating a specific subsystem may require fictitious weather data and changes to the software source code (Judkoff et al. 2008).

## D.2 Implement Possible Model and Solution Processes

Isolating the potential issue to a subsystem may not always be possible, because the software may not model the related building components, occupant behavior, or site conditions. For example, if the potential issue is "comfort effect from natural ventilation on occupants," there may be no model for determining air velocity profiles in the space, and no model for the effect of air velocity on occupant comfort. If no subsystem can be isolated, or if the models in the subsystem are not adequate to investigate the potential issue, it may be necessary to modify existing models or develop new models and solution processes.

## D.3 Verify Against Analytical or Numerical Solution Processes

The first step in evaluating a subsystem is to verify that the solution processes are implemented correctly. Under simple input conditions, it may be possible to develop an analytical solution (solve the governing equations exactly) or create a verified numerical solution[16] for the isolated subsystem. For example, one could verify the solution process for a 1-D wall heat transfer subsystem by comparing predictions of the isolated subsystem to the steady-state analytical solution for 1-D heat transfer through a wall. The analytical and verified numerical solutions are developed based on the physical models in the software, so a high level of agreement is expected in these comparisons.

Agreement with one analytical solution does not ensure the solution process is implemented correctly; the solution process should be tested versus analytical and verified numerical solutions implemented over a range of parametric variations. Continuing with the example above, after confirming agreement with the steady-state solution, it would be important to test the solution process under dynamic conditions (e.g., input dynamic surface temperatures sufficiently simple that an analytical solution exists, such as a step function or sinusoid).

## D.4 Correct Solution Processes

If the software predictions do not agree with the analytical or verified numerical solutions, the solution processes must be corrected. Corrections may involve bug fixes, alternative spatial and temporal discretization, or alternative numerical methods. As an example, Figure 9 shows actual results of applying subprocesses D.1–D.4 for the conduction finite difference heat transfer algorithm in EnergyPlus; the model was isolated in EnergyPlus v5 ("before") for heat transfer

---

[16] A verified numerical solution is generated using another analysis tool in which the isolated subsystem can be replicated and solved under controlled input conditions. For example, many numerical equation solver programs are available to modelers. The mathematical equations corresponding to models in the isolated subsystem can be entered into these tools and solved for specific input conditions. One must be confident that the numerical solvers in the secondary tool have been previously-verified, and that spatial and temporal discretization and convergence criteria are sufficient for appropriate accuracy. For more details, see Neymark and Judkoff (2009b).

through external walls, tested against numerous analytical and verified numerical solutions, and then errors in the code were corrected and included in the EnergyPlus v6 release ("after") (adapted from Tabares-Velasco and Griffith [2011]).

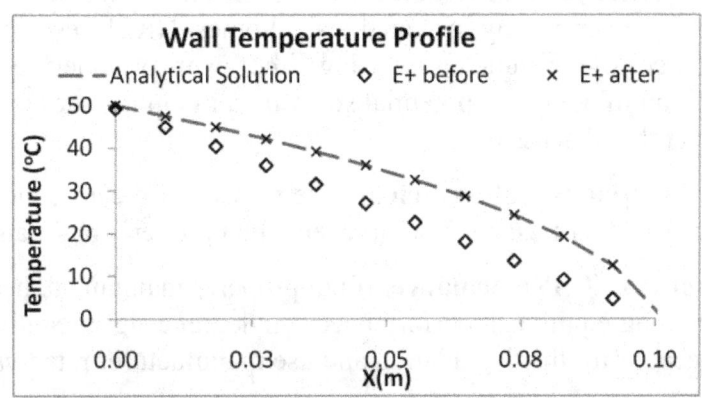

**Figure 9. Example results for model isolation, verification, and solution process correction (adapted from Tabares-Velasco and Griffith [2011])**

## D.5   Validate Against Empirical Data

If the software shows good agreement with analytical and verified numerical solutions (Section D.3), the solution processes for the subsystem have been verified. The next step is to validate the models in the subsystem against empirical data from controlled laboratory experiments or field tests. The experiments or field tests must be carefully designed to control or measure all input parameters and measure all output parameters necessary to validate the subsystem (see Neymark et al. 2005). The general approach is to simulate the experiment with the software using measured input values and compare software predictions with measured output values. Measurement uncertainty and error propagation techniques should be used to understand the uncertainty in measured and predicted results. If the results of the software do not agree within the extent of experimental uncertainty, the isolated subsystem does not completely represent the physical phenomena in the experiment and must be modified or redeveloped (see Section D.2).

## D.5   Document and Implement Solution

If the model agrees within the experimental uncertainty of the empirical data, its solution is considered validated for the range of conditions investigated in the controlled experiments or field tests. Models with validated solutions are documented and published individually. The impact of not adopting the improvement is also estimated and described in the documentation to help software developers prioritize possible software improvements. Improvements are submitted for inclusion in a future release of the reference software.

# Appendix E    Initial List of Potential Software and Input Issues

At the 2005 Affordable Comfort Conference, Dave Roberts and Michael Blasnik (2005) hosted a roundtable session titled *Real vs. Rated Energy Use*. The panel of experienced building scientists and the audience were invited to brainstorm ideas as to why simulation software might overpredict energy use in older, poorly insulated leaky homes. NREL researchers have also discussed potential sources of inaccuracies (causing either over- or underprediction) for many years. The following is an initial list of potential software and input issues based on the roundtable session and other discussions:

- **HVAC System Capacity Not Sufficient.** The furnace or the air conditioner may not be able to maintain the thermostat set point, even when properly maintained.

- **Degradation of HVAC Performance.** If not properly maintained, heating, ventilating and air conditioning equipment performance can degrade significantly with time. Many tools do not account for the degradation and use manufacturer rated values for system performance.

- **Temperature Variation Room-to-Room.** Often air temperature in the entire home is assumed to equal thermostat set point. Temperatures in remote rooms may deviate significantly from thermostat set point, especially if occupants close supply registers.

- **Temperature Variation Within One Room.** Most simulation tools assume uniform air temperature throughout a room. This may not be accurate. Temperatures throughout the room may be stratified, especially in some older homes where air distribution systems do not adequately mix the air.

- **Shielded Walls, Floors, and Ceilings.** Furniture, wall hangings, cabinets, and closets provide additional insulation for exterior surfaces that usually is not accounted for in building models.

- **Performance of Uninsulated Assemblies.** Current assumptions about the effective thermal resistance of air film coefficients and air movement through uninsulated assembly cavities could be inaccurate. Errors in film coefficients and air-gap thermal resistances, although small in terms of absolute thermal resistance, can lead to significant errors in the prediction of heat transfer through poorly insulated assemblies. This is because film coefficient and air-gap thermal resistances represent a larger percentage of the overall assembly thermal resistance than they do in high-efficiency assemblies.

- **Infiltration air heat recovery.** Ambient air entering the building through walls and other enclosure assemblies exchanges heat with warmer or cooler assembly components, potentially moderating the thermal impact of infiltration.

- **Thermostat settings.** There is significant uncertainty in the values used for thermostat settings. For an asset analysis, "typical" thermostat set point schedules may not reflect average settings and therefore may not yield average energy use predictions. For an operational analysis, the set point schedules may not reflect the actual behavior of the occupants, who can change thermostat settings manually or reprogram thermostat settings throughout the year.

- **General infiltration modeling.** Many infiltration models use information from a single blower door measurement, wind speeds and outdoor temperatures from weather files, and information about the shielding of a building to predict hourly air infiltration rates. In the past researchers observed over-prediction by commonly-used infiltration models (Francisco and Palmiter 1996), and more research is needed to evaluate the accuracy of current infiltration models.

- **Foundation heat loss.** Heat loss through building foundations is a 3-D phenomenon and has dynamic effects on time scales from one hour to several years. Current practice in modeling foundation heat loss is to approximate the problem as 1-D heat transfer. For uninsulated foundations, heat loss is proportional to the conductivity of the surrounding soil, a thermal property that is difficult to quantify and can vary by a factor of 5 (Sterling et al. 1993), depending on moisture content and type.

- **Insulating effect of snow cover on roofs.** Snow can provide additional insulating value for roof assemblies. Some software may not account for this effect.

- **Building solar shading.** Homes are shaded by immediate neighbors and by mature trees. Some simple tools cannot account for these effects. Buildings are sometimes self-shading, and this may not be modeled. Windows have many complex geometric features that reduce the gain of solar radiation, and these features may not be modeled well.

- **Differences in occupant behavior due to building performance.** Occupant behavior may depend on the efficiency characteristics of the home. For example, an occupant may be more likely to turn down the thermostat setting in response to high heating cost in an older home than in a new, well-insulated, tight home. Alternatively, they may use less-efficient thermostat settings to overcome uncomfortable radiant temperatures and drafts in older homes.

- **Wind shielding mischaracterized.** Wind is a major driver of whole-house infiltration and exterior convective heat loss coefficients. Underestimation of wind shielding will typically lead to overprediction of infiltration rates and heat loss through exterior surfaces, especially on poorly insulated surfaces where the film coefficients are a large fraction of total thermal resistance.

- **Insulation performance different than tested values.** Insulation ratings are determined from tests conducted at near-room temperature. The conductivity of these materials is generally lower under colder conditions.

- **Insect screens or other window treatments.** Insect screens on windows reduce solar gain and provide protection from wind-induced convection loss. These are generally ignored when simulating performance of homes. The impact will be more pronounced on single-pane, clear glazing than modern, double-pane, low-e, low solar gain windows.

- **Surface heat transfer algorithms.** Energy use and savings predictions can be very sensitive to surface heat transfer algorithms and assumed film coefficient values when considering homes and retrofit measures involving poorly insulated pre-retrofit building components (e.g., single-pane windows, uninsulated walls). Research is needed to understand these sensitivity issues, evaluate surface heat transfer algorithms in the context of existing homes, and provide updated recommendations for average film coefficient values used in simplified modeling approaches.

- **Multidimensional heat transfer.** Heat transfer through building enclosures is typically modeled as a 1-D process. In reality, 3-D effects may cause significant differences between actual and modeled heat flows.

- **Secondary fuel sources.** Secondary fuel sources such as fireplaces are highly variable and may account for significant discrepancies between predicted and metered energy use.